The Electronic Link to Retail Also Known as EDI (Electronic Data Interchange)

by
Alan Castro

authorHOUSE™

1663 LIBERTY DRIVE, SUITE 200
BLOOMINGTON, INDIANA 47403
(800) 839-8640
WWW.AUTHORHOUSE.COM

First published by AuthorHouse 09/08/05

ISBN: 1-4208-7597-3 (sc)

Printed in the United States of America
Bloomington, Indiana

This book is printed on acid-free paper.

Table of Contents

1

Introduction to EDI

Electronic Data Interchange has been used as a tool for stores, also known as the distribution center or the processing center, to reduce the paper flow to its vendors and also from its vendors. As the vendors get to take advantage of the technology behind the concept of the Electronic Data Interchange, also know as EDI. EDI has now been in place for many years and when used properly, has had the effect of moving production much faster and much more accurately.

EDI has moved forward with set documents which allow stores and vendors to understand what information will be needed to have a complete order cycle completed.

After the order has been placed, the vendor has returned an acknowledgement by the vendor so the store knows that the orders have been received. Invoices will have been sent for payment, and then an acknowledgement has been sent by the store to say that their invoices have been received. Next the Advance Ship Notice will be sent usually with a Window of up to seven days before the shipment is picked up by a trucking company or sent out by the vendors own trucks. The window of days can vary from store to store. As the Advance Ship Notice is received by the store, the store in return will send an acknowledgement to say that they have received the advance ship notice.

The acknowledgement, also known as a Functional Acknowledgement, plays a major part for all sides involved because there usually will be no other communications between both sides. The flow of EDI was setup to always use the given documents as the tool of communications. As in any situation, you can encounter a piece of information that has been submitted wrong. After all the factor of human error plays a small part and this can happen.

As we move on into the setup of documents, you will see how the documents are all relative and how they all relate to a specific company and a specific vendor. All communication setups are unique to the store as well as it is to the vendor. As we move in to the setup process, it will be clear how the importance of a proper setup comes into play. As most stores require a test to be sure that the communications is properly setup, the documents are fully mapped and both sides know what to expect once the process of EDI is in production.

2

The Setup of an EDI Partnership

Taking on the task of setting up a vendor on EDI is a little more involved than just getting confirmation from the store that they will start to transmit purchase orders to the vendor and receive invoices and advance ship notices from the vendor.

Each store will have a set pattern of rules to follow. The first is to start the registration and fill out the startup forms. The store will tell you how to access these forms on the web site. These forms will be filled out on the web and submitted thru the website. These forms will ask you for the vendor information. This will include billing and EDI contacts that will assist them in the process of moving forward. All forms will be found at the web sites that have been setup by the store. These forms will call the setup a partnership between the store and vendor. At this time you will determine what your ids will be. One id that is referred to is a sender/receiver id. All sender/receiver ids' have a qualifier assigned to it. How this works is that a Duns number can be used, and a 08 qualifier will be assigned to it. You can use a telephone number and a qualifier of 12 will be assigned to it. Some stores will use an identification number that will have a 01 qualifier assigned to it. It is important that you know that once you register this sender/receiver id and qualifier with the network, that you will not be able to change it. Although it is not impossible to change, it can be costly and time consuming. All changes would have to be recorded on the store end as well as the network end.

As an example: 08:05523879
12:2125558900
01:6111570035

When the forms have been completed, and submitted electronically, the next step is the electronic partnership. The electronic partnership is at the level of the network. The network, which there are several of them, such as GE (Global Exchange), Sterling, Inovic (Formally QRS), is where you will have an electronic mailbox. This mailbox is setup for a specific vendor, has a user id assigned to it, an account # and password will also be assigned. The user id, account #, and password will only be given to an account manager, company owner or the person setting up your EDI for you. Anytime that you call the network with a question and or problem, you will be asked to verify this information. That is how your account is accessed. Most of the time you can access your account on their web-site using the above information. Each network has its own site that they will tell you about and how to access it.

When you are setting up an individual store, after the setup forms have been completed, and the network has been completed, it is now time to meet the demands of that partnership. The first requirement would be the testing of a purchase order. Once you have received the test purchase order, you will be able to accept the fact that your setups are finished and have been setup properly.

Upon receiving a test, you will send a functional acknowledgement back to them in test, to let them know that you have received their test purchase order. In most cases, you will be required to turn it around and send it back to them as an invoice in test. In most cases, the store will also ask you for the paper invoice to compare the information that was sent electronically and the original paper invoice. It is also important to know that the site that you got your setup forms from, you will also be able to get the mapping of documents. As the documents follow a pattern, some stores have some differences to them. These differences are parallel to how they handled their documents from the start. In the case of the Advance Ship Notice, also known as an ASN, they will ask you for a test UCC128 label.

3

Creating and Receiving a Purchase Order

Your system will dial into a phone number that will be given to you by the given network that you have chosen to use for your EDI processing. This will start the communications process and transmission process of your orders coming into your system. All this is done with computer modems on each side. Your modem will dial to the given number on the network side and will connect to start the transmitting of data on each side. As your modem connects with the other modem on the network side, they will begin by sending and receiving signals to complete the transmission cycle. When the transmission has completed and your modem regains the idle stage, this will indicate that the transmission has finished. At this time you will need to extract the information from the EDI documents that were sent to you. Remember that some of the information given to you is only needed for the purpose of defining the document formats that were sent to you to receive the purchase order. Let's define the documents and their formats so you will see how it all ties together and how you will be able to bring it into your system to process as an open order.

ISA is the start of your document (also known as an envelope or interchange), and has its ending document partner IEA (interchange control trailer). The information in the ISA format will consist of your ids. Your sender/receiver id and the partnerships sender/receiver's id's will be in this segment. The date and time that the order was submitted for will be in this segment of the transmission, the time that the order was submitted for, is in the segment. A code of a "P" for a production transmission will represent a production transmission. This will indicate if the order was transmitted for the purpose of a real purchase order or a code of a "T" for the purpose of just testing, in which case the purchase order is usually a duplicate order that you may have processed in the passed. You usually are aware of what stage you are in when transmitting. In the ISA is a control # that is sent to you from your partnership and is usually a sequential number. This number is very important because if there is a purchase order that may not have been received or transmitted, this number will serve as a tracking number.

The IEA segment will match this number. This is a reference to the communications cycle that the envelope is complete and the cycle is done. If you were to send two different numbers, the ISA control # and the IEA control # not matching, the network will not process it on either side because it would indicate that the cycle may not have been completed and the wrong numbers were submitted. The GS segment will have a functional Id of PO in it to represent the document as a Purchase Order.

ST (start) segment is the internal start of your information that has a counter in it that will represent the portion of the purchase order that will be processed on its own. As an example, The ST for 0001 may represent a certain purchase order and ST 0002 may represent the next purchase order. Before the purchase order changes, you will have an SE (segment ending) segment that will match the ST number before it changes to the next ST segment. These control #'s are always used as tracking numbers and verification #'s for you and your partnership. There is always a chance that when a purchase order is submitted that a key in error will occur. If you have the information coming directly from your system, then the object of error will be close to none, but if there is keying in taken place, then the error factor can be higher. In these cases the control #'s play a very important purpose in fixing this error.

BEG segment is the beginning of the purchase order transmission that actually pertains to the transmission of the purchase order information. BEG (beginning) is the segment that you will find the purchase order date, and the purchase order number. In this segment is a code that will denote that the order is an Original Order, blanket order, Stand Alone Order, Release Blanket Order. This code is very important. After this segment you may find the NTE (note or special instructions). After this segment you will find the REF segment (reference) that will define the department number and the internal vendor number. DTM segment (date movement) will define your start ship dates and your cancel dates. These dates are the window for the purchase order to be processed and shipped. If passed the issued cancelled date, the purchase order would need to be extended by your buyer or can be open for cancellation. Some partnerships include MSG segments (message) that can give you information about routing and information about shipping your order or the NTE segment which will also have messages about shipping your order in it.

Your next segment is the main portion of the transmission. This segment is the PO1 Segment (purchase order 1). This segment will give you a line number, the quantity that will relate to the UPC code that is given, the selling price of the UPC code (style or item that is used in this order), and in many cases the partnership's SKU number. The PO1 segment will also give you the unit of measure. Such as EA for eaches, DZ for dozens, CA for cases, and AS for assorted. The next segment is the SDQ segment. This segment is not always used the this document. If the SDQ segment is used, it will have the store numbers and the amount of the quantity ordered for each store. There can be a number of stores and quantities in this segment. The next segment is the PID segment (product identification description). This segment is a freeform segment that will give you the description of the styles or items that has been stated in the PO1 segment. Each PID segment will have an "F" before the description to denote that the description is freeform. The CTP (pricing information) segment will have the unit of measure in it. If the purchase order is packed in 12, 24, and 36 and so on, this segment will show it. In this segment there will also be the style or item cost in it. As an example, The PO1, PID and PO4 segments can appear multiple times. This is determined by what has been ordered in the purchase order. After all the details of this purchase order has been entered. Closing in on the end of the purchase order, you have a CTT segment (counter total) that will have a count of all the lines of purchase order detail in it.

The following is an example of a Purchase Order:

ISA **00** --*S/R--------- --*S/R ID--------- *----------*TIM

E*US Code*Version/Release*Interchange Id*Ack Request*Prod*>*

GS*PO*-----------*------------*Date*Time*Assigned*AnsiX12*Version

ST*Document Number*Transaction Set Number

BEG*Original Code*PO Type Code*PO#*PO Date

REF*DP*Department

NTE*Delivery Code*Special Instructions

DTM*001*Start Ship Date

DTM*010*Cancel Dates

N1*ST*92*Store #

N3*Address

N4*City*State*Zip

PO1*Line #*Quantity*UOM*Unit Price**ID*Product #

SDQ*Store Number*Quantity being ordered for that store (there can be multiple stores and quantity amounts in this segment

CTP*Class Code*Price Code*Retail Price or Cost

PID*Item Code*Product Code***Description

SDQ*UOM*ID Code*Store *Quantity

CTT*Total of Line Numbers

SE*Transaction Set Count*Transaction Set Number (From ST Segment)

GE*Number of Included Segments*Interchange Control from GS Segment)

IEA*Number Of GS Segments*ISA Control #

All of the segments end with a binary delimiter. This is a character that is created by a binary chart. BINARY delimiters is the process of opening the computer characters and closing computer characters. The * is also a binary code for the field separators or delimiter. This can vary from store to store or partnership to partnership. This is also written in binary, from the binary charts.

According to your partnership, this document can have some small differences to it. According to what version you are using, this document can have small changes to it.

The Functional Acknowledgement is the document that will Notify your partnership that the purchase order has been received and has been accepted. In this document there is a segment that will state how many purchase orders have been received.

An Example of a Functional Acknowledgement:

ISA*00* *00* QL*S/R ID QL*S/R ID *Date *Time*US Code*Version*Control #*0*P*>

GS*FA*S/R ID*S/R ID*Date*Time*Data Control#*ACK Request*Version

ST*Document number*Transaction Control Number

AK1*PO (What this FA is for)*PO Number

AK2*850 (Document number this is for)

AK5*A (Approved or E (Error) or R (Rejected))

AK9*A (Approval Code)*Number of PO's*Number of PO's Sent*Number of PO's Accepted

SE*Number of Included Segments*Control Number

GE*Number of Included Transaction Sets*Data Interchange Control Number

IEA*Number of Included Groups*Interchange Control Number

The ISA Segment will be the same format as the Purchase Order. This is also the same for the GS/SE segments, and the IEA Segment. In the GS segment, you will find an FA, as the functional Id. This is to show you that the document is a Functional Acknowledgement. This document has a record control delimiter at the end of each line. This record control delimiter is a character that is represented in the form of a binary code. The binary codes are written in conjunction to the binary charts. This document will also be used by your partnership to let you know that your invoices or advance ship notice was accepted or rejected when you have sent them. In the AK1 segment the document number will be denoted to which document this is being used for. The 850document is the purchase order. The 997 is the functional acknowledgement. The 810 is your invoices. And the 856 is your advance ship notice.

4

Creating an Electronic Catalog

The electronic catalog is where you would load up your product line of styles and UPC codes for your partnerships to be able to access and review while using it to write your orders. This document is also referred to as an 832 document. This document will breakdown your products or styles into the detail of each product or style, including the product or style number, color, sizes and descriptions. This document will use the NRT codes. These NRT codes are the National Retail Codes that have been setup to define the product to a little further detail and also categorize them. These codes will represent their color and size. As the setup of this code are all different for each category, it would be better to table them up and just cross reference them when needed.

For color codes, you will take the major color and break it down as if you were using shades of that color. As for the sizes, they are broken down by Mens and Womens sizes. A book has been put out to breakdown these categories.

This document relies on the the date that you have loaded your catalog, to represent the product or style. This date will be read by the partnership that is using it, and will know if it is a current or discontinued product or style number.

This document can be transmitted to the networks, such as the GE (Global Exchange) or Inovic (Formally QRS) network. Some partnerships are able to accept their own 832 documents and some others will only accept 832 documents thru the network. As a note to remember, some partnerships take twelve to twenty four hours for their systems to load up the information that you have transmitted to them.

Your UPC Code (Universal Product Code) will consist of a registered number that has been assigned by the Universal Council Code. This number will represent your company because it is registered to your company name. This number is a six digit number that will be represented in your UPC code. The next five digits will be a sequential number or a number that will represent your product or style number, could have a number to represent a color and or a size in it. The last digit is a check digit that is calculated by the first eleven digits. You can add up the odd numbers, take the total and multiply by three. Next you will add the even numbers. Add the calculated odd numbers to the total of the even numbers, then subtract from the nearest 10th, and the number will be your check digit. This number will be used to create a Bar Code. This is a number that will get scanned and will be checked against the catalog that you have loaded. This also acts as a validation for this product or style UPC Code.

As an example, if your company has printed a UPC Code on your products ticket, and it was not put on the partnerships catalog, when it is scanned at the register, it will come up and state that it is an unknown UPC Code. So it is important that the Electronic Catalog is entered and loaded properly and is always current for the partnership and your benefit.

The following is an example of an 832 Catalog Document:

ISA*00* *00* QL*S/R ID QL*S/R ID *Date *Tim e*US Code*Version*Control #*0*P*>

GS*SC*S/R ID*S/R ID*Date*Time*Data Control#*ACK Request*Version

ST*Document number*Transaction Control Number

BCT*RC (Resale Catalog)*Id Number pertaining to your company (usually your phone number)*000 (Selection Code pertaining to the category that your merchandise would be in, such as 001 will equal blouses, 002 will equal sweaters, 003 will equal jackets)*02 (Code to represent the action you are taking with this document. Such as 02=ADD a record, 04=Change a record, 03=Delete a record)

DTM*043 (Publish Date), 092 (Booking Date), 018 (Available Date), 036 (Discontinue Date)*CCYYMMDD (Date in Century, Year, Month, Day)

N1*CT (country of origin code)*Country (the country you are in)

N1*SF (company that you will be sending the catalog from, in a free form name)*Company Name (the company name that you are sending this catalog from)*Buyers Assigned Code (91)*Vendor UPC Code

N2*Address of the company that is sending this catalog, this is free form

N3*Continuing the free form address

N4*City, State, Country, and Zip of the company that is sending this catalog

LIN*1 (number of detail segments)*UP (The code to represent the next field which will be the UPC Code)*UPC Code (The UPC Code that will represent the Item number or style number that you want to add, change, or delete onto the catalog)*VN (the code that will represent the next field, the vendors catalog number, item number or style number)*item number or style number (the item number or style number that is cross referenced to your UPC Code)*CM (code to represent the color code. This will also be an NRT Number)*Color Code (this is the NRT Number that will represent the color of the item or style that you are sending)*SM (code to represent the size of the item or style that you are sending/ This will also be an NRT Number)

DTM*043 (Publish Date) or 092 (Booking Date), 018 (Available Date), 036 (Discontinue Date)*CCYYMMDD (Date in Century, Year, Month, Day)

CTB*OR (ordering)**70 (quantity qualifier-minimum 57, maximum 70)*

70 (minimum or maximum amount of units)

PID*F (code to represent the description in freeform)*08 (code to represent the description of the item or style)***the description of the item or style in freeform

PID*F (code to represent the size in free form)*74 (code to represent the name of the size)***the name of the size (the free form size of the item or style)

PID*F (code to represent that the color name that will be in free form)*73 (code to represent the color name of the item or style)***the color name of the item or style)

PID*S (structured)*VI (vics)*RY (merchandising-can be reordered)
        CL (collateral)
        GW (a gift with purchasing)
        PP (UPC Prepack)
        PW (Purchase with purchase)
        RN (merchandise cannot be reordered)

PID*F (code for free form name to have the catalog represented)

*TRN (trade name)*Name of item line or the name of the style line

PID*F (code for free form)*92 (represents the next field which will be the fabric description)*fabric description (free form description of the type of fabric used or the type of material used to produce the item)

CTP*RS (resale)*MSR*0 (reorder point)**EA (unit of measure)

(DTM to the last PID will loop for every item or style that is sent to the catalog)

CTT*number of line items that are being sent in this transaction

SE*Transaction Set Count*Transaction Set Number (From ST Segment)

GE*Number of Included Segments*Interchange Control From GS Segment)

IEA*Number Of GS Segments*ISA Control

Notes to Remember

All of the segments end with a binary delimiter. This is a character that is created by a binary chart. BINARY delimiters is the process of opening the computer characters and closing computer characters.

As an example the * is also a binary code for the field separators or delimiter. This can vary from store to store or partnership to partnership. This is also written in binary, from the binary charts.

5

Creating and Sending Invoices Thru EDI

Creating and sending invoices is a very important step. As everyone wants to paid, this is the way everyone can get paid in a faster and more accurate manner. The 810 document is the invoice document. In most cases, the information will be transferred from your system to the EDI invoicing system. This is the step that allows you to transmit your invoices instead of mailing them. When transmitted to your partnership, they will be loaded into their system and will start the aging process that will be in conjunction to your terms that were setup by you and your partnership. The partnership will return a functional acknowledgement to state that it went over and was received by them with their approval.

The documents will once again begin with the same segments as all of the other documents. The ISA will start the segments and will have the Sender/Receiver Id's of you and your partnership. Your Partnerships Sender/Receiver can be different for your purchase order document. These Id's are assigned by your partnership and will also be given to you upon setting up this document. This segment will also follow the other documents, with the IEA as its partner. As most partnerships have developed on their web-site, are the setup and start up forms that you can fill out and submit thru their web-site. When your confirmation is returned, your Id's are usually included in the confirmation so when you start to setup your system, you can put in the proper Id's to insure that the transmission will be received by the correct department. The GS segment will also follow the other documents except for the beginning Id. In this document, the functional Id will be IN for invoice. This document will also have the GE segment as its partner. The next segment is the ST (Start of the Transaction Set). This segment will also follow the other documents, with the SE as its partner.

The BIG segment is the beginning segment that will start the process of each invoice that will be sent. In this document the BIG segment will be to represent each invoice sent to your partnership. In this segment, the invoice starts to show its detail. The invoice number, invoice date and the purchase order number and purchase order date. The following segments will continue to show the detail of the invoice, such as REF segment which will represent the department number and the N1, N2 and N4 segments the will tell your partnership the name, address, city, state and zip code of the store that you had shipped your merchandise to. The ITD segment will define the terms of the invoice. In this segment you will find the most partnerships have defined some fields with standard codes. In the first field, the terms are usually defined as: 01: Basic

> 02: EOM
> 03: Fixed Date
> 04: 10 Days after EOM

The third field is the percentage of the discount, if there is a percentage then You would enter the number as a whole to represent the discounted amount. If field three has a discount in it, then you would need to put in the discount due date in field four. The next fields are optional. They are the Terms Net Due Date and the Terms Net Days. Usually if field four has been entered, you would have field eight filled in with the terms discount amount. Other fields that are usually not required are the next fields until you get to field twelve which is the description field. Fields nine thru eleven are not used that much because they refer to the deferred discounts on an invoice. This is not often used. Fields thirteen thru fifteen are not often used either. These fields refer to more specific information such as the month and payment method codes. Then the next segment to follow would be the DTM, date segment. Using the ship date code of "011", you would at this time have the ship date put out. The ship date will denote to your partnership that the Advanced Ship Notice has been sent and can be used for comparison. The ship date will also be compared to the date in the GS segment to show your partnership that you have not sent this invoice before you have actually ship the goods.

In some cases you will need to use a code of "067" that will tell your partnership what the cancel date is. The next segment usually is the FOB segment. This segment will specify the transportation instructions relating to the shipment. In the first field the codes for freight payment will be used. The codes are CC for collect and PP for prepayment. In some cases the second field would be used to denote the destination for shipping or the origin (shipping point). These codes are usually DE or OR. The segments from BIG thru the FOB segments are known as the header information and usually exist at the top of your invoice. The next segment to follow would be the IT1 segment. This segment would specify the most frequently used line item data for your invoice. This segment will repeat itself for the same number of item lines that you have in your invoice.

In this segment there is a line number to show your partnership that you have sent the same number of items that you have shipped. Next field, field two, would have the quantity shipped for that item. In field three, the unit of measure would be represent by a code of how you have shipped the merchandise. The codes that are commonly used are the EA code for eaches, the code CA for cases, the code DZ for dozens, and the code AS for assorted. The forth field would be used for the unit price. The sixth field is the code for the partnerships catalog style or item. In the seventh field would be the partnerships catalog number. This number could be the item number used by your partnership as their inventory number or SKU number they use on their system. In the eighth field would be the catalog code to represent your inventory or system style or item number. This code is usually represented by a "VA". The ninth field would actually be the style or item number sold on this invoice. The tenth field would be the code to represent the UPC code that you have representing the style or item number sold on this invoice. This code is a UP. In the eleventh field would be the UPC code that was used to represent the style or item sold on this invoice.

In some cases, the actual style or item number may not be needed to be sent Back to them. In some cases, their SKU number or their catalog number may not be needed to be sent back. In almost all cases the UPC code is needed in the invoice. For every IT1 segment, you will need in most cases the PID segment as a following segment. This segment is the product or item description. In the first field, a code of an "F" would be needed to represent the fact that the description is a free form description. Usually the only other field that is used in this segment is the fifth field. This field is the style or item description. In some cases, the SLN segment will be teamed up with the IT1 and the PID segment. The SLN segment is the product sub line detail item data. The SLN segment usually only utilizes the fourth and fifth fields. The fourth field being the quantity and fifth field being the unit of measure. If the partnership was to utilized the SLN segment, they might also use the sixth field as the unit price, the ninth field as the catalog number code (CB), and the tenth field as the actual partnership catalog number. The eleventh field would be the vendors item code (VA). The twelfth field would actually be the vendors style or item number. The thirteenth field would be the code for the UPC code. The code is represented by a UP. The fourteenth field is the actual UPC code. The fifteenth code is the vendor code that will represent the vendor color. This code is a VE. The sixteenth field would be the actual color name. The seventeenth field is the code for the vendor size. This code is a SZ. The eighteenth field is the actual size of the style or product. In the next segment, you move to the summary part of the invoice. The next segment would be the TDS segment. This segment will define the total of the invoice. This total will include any freight charges, discount amounts and allowances. The next segment is the CAD segment. This segment is to specify the transportation details for this invoice. In this segment, the only fields that are used are the fourth field, which is the SCAC code.

The SCAC codes are the standard alpha codes that represent the trucking company. The sixth field will be represented with a code to show the order status of the invoice. The codes are CC for the order that was shipped complete or the code PR for the order that was partially shipped. In the seventh field will be the Bill Of Lading code. This code is represented with a BM. The eighth field is the Bill Of Lading number. The next segment is the SAC segment. This is the summary segment that is usually used if there are discounts, allowances and Freight. The next segment is the ISS. The ISS segment is the invoice shipment summary segment that will summarize the total number of cartons and the total amount of weight. The final four segments are the CTT segment (control counter), the total number of detail lines, the SE segment is the partner of the ST segment, which is the transaction set trailer, the GE segment is the partner of the GS control number, which is the group control number, and the IEA segment is the partner of the ISA control number, interchange control trailer.

Notes to Remember

The invoice document, also known as the 810 document, has a field separator which is created by binary coding, also known as a delimiter. Each segment has a record control delimiter, which is also created by binary coding.

Always keep in mind that your partnership may post in their documentation what delimiters that they want you to use. This means that you will have to include these delimiters in your document so their system will be able to read the document.

The following is an example on an Invoice Document:

ISA*00* *00* QL*S/R ID   QL*S/R ID   *Date *Time*US Code*Version*Control #*0*P*>

GS*FA*S/R   ID*S/R   ID*Date*Time*Data   Control#*ACK Request*Version

ST*Document number*Transaction Control Number

BIG*Invoice Date*Invoice Number*Purchase Order Date*Purchase Order Number

REF*Reference Qualifier (such as DP)*Reference Number (such as department number)

N1*Identifier Code (Such as ST=ship-to, Z7=mark for store, SF=ship from)**Identification Qualifier (such as 92, you are going to use the store number, 9=Duns number, 91=Phone number)*Identification code (such as the store number, duns number, or phone number)

N3*Free Form Address

N4*City*State*Zip

ITD*Terms Code Type (01=Basic, 02=EOM, 03=Fixed Date, 12=10 Days

After EOM)**Terms Discount Percentage*Term Discount Due Date (If The Percentage Exist, Then This Must Be Filled Out)*Terms Net Due Date (Only If Discount)*Terms Net Days (Only If No Discount)*Terms Discount Amount (Only If Discount Exist)****Description (Free Form)

DTM*011(Shipped Code)*Shipped Date

FOB*CC(Collect), PP(Prepaid)*Location Qualifier (DE=Destination of Shipping, OR=Origin (Shipping Point)(Location Is an Optional Field)

IT1*Line Number*Quantity Invoiced*Unit of Measure (EA=Each, CA=Cases, DZ=Dozens)*Unit Price**CB (Code for Buyers Catalog)*Sku Number or Buyers Style or Item Number*VA (Code for Vendors Style or Item Number)*Vendors Style or Item Number*UP (Code for UPC Number)*UPC Number

PID*Item Description Code (F=Free Form Style or Item Description)****

Style or Item Description (This Segment Is Optional With Some Partnerships)

SLN*Line Count***Quantity*Unit of Measure (Most Partnerships Only Require These Fields)

TDS*Amount of Invoice (This Amount Will Include Charges, Less Allowances, And Freight)

CAD****SCAC Code (Trucking 4 Digit Alpha Code)**Shipment or Order Status Code (CC=Shipped Complete, PR=Partial Shipped)*BM (Code for Bill Of Lading Number)*Bill Of Lading Number

SAC*Allowance or Charge Code A=Allowance, C=Charge)*Code of Action (C000=Defective Allowance, C540=Early Buy Allowance, D200=Freight Charge to Destination, D240=Freight, E740=New Store Allowance, F050=Miscellaneous Charge, H000=Special Allowance, I060=Ticketing Allowance, I570=Warehouse Allowance)***Total Dollar Amount of Allowance or Charge (This Segment Is Optional If You Do Not Meet Any of These Requirements)

ISS*Number of Cartons Shipped*Unit of Measure Code (CA=Cases)*

Weight (Total Weight Shipped)*Unit of Measure Code (LB=Pounds)

CTT*Total of Line Numbers

SE*Transaction Set Count*Transaction Set Number (From ST Segment)

GE*Number of Included Segments*Interchange Control from GS Segment)

IEA*Number Of GS Segments*ISA Control #

It is always important to remember that each partnership was vary due to the requirements of that store.

6

Creating and Sending an Advanced Ship Notice Thru EDI

Creating an Advanced Ship Notice, also known as an ASN (856 Document). This is the document that actually defines the shipment that is due to arrive at the distribution center or in some cases, directly to the stores. The ASN in many cases has the same components as other documents but also has many variations in the document. This is related to the partnership that will have their own set rules on the contents of the ASN, but more important is that the partnerships have their own rules of how they want this document to get created.

Let's explore some of the rules set by your partnerships. In some cases you can create an ASN document up to seven days before the shipment will arrive. The ASN will remain on your partnerships system and will not get used until the shipment arrives and starts to get scanned in. It is important that the ASN is on their system before the shipment arrives. That is why it is called the Advanced Ship Notice. Other Rules that exist with partnerships are trucking documents (753 and 754 Documents must be sent first, within a forty eight hour period), must be sent, this document would be needed to be answered first. The Routing Document (753) is the document that will relate the trucking information to them, such as the amount of cartons, total weight, total cubic feet, and exact ship date and invoice number. When this is sent, this will detail the shipment to the partnership so that they know that the truck that will arrive to pickup the merchandise is the correct truck.

In return the 754 document will be returned to give you back the trucking information. In most cases, the Bill Of Lading number should be used in your ASN, when sending it. In some cases your partnership will give you twelve hours to create the ASN and send it over to them. In some cases your partnership can give you up to three days to get your ASN sent to them before the shipment arrives. The rules from your partnerships are usually posted on your partnerships Web-Site.

The ASN is actually a two step process which will require the ASN being transmitted and a UCC 128 label being created to represent the UCC 128 carton number.

The ASN begins with same segments as the other documents with the ISA envelope opener. The ISA will carry the Sender/Receiver number of your company and the Sender/Receiver number of your partnership. This information would be posted on your partnerships Web-Site. In many cases, you will find that the Sender/Receiver ids will not be the same as the other documents. This allows the partnership to have the extra separation when processing the document on their system. The next segment is the GS segment. The GS segment is the same as the other documents except the Sender/Receiver numbers again will vary from partnership to partnership. This GS segment will have a code of SH in it to show that this is the ASN document. The next segment is the ST segment. This segment will start to denote that the document is an 856 document along with a sequential number. The next segment is the first HL segment. This segment will start the beginning of the shipment level data. There will be only one order level data area for each different order in the shipment.

There will be one carton level data area for each carton in the order. This HL segment will be identified with an "S", to show that this is in the shipment document and a sequential number to identify this transaction set. The REF segment will be used to start telling your partnership about shipment information. In this segment we will reference the Bill Of Lading number and in some cases there will be another REF segment to reference the department.

The N1 segment will follow to state the ship-to information. The N1 segment will require the name of the store that you are shipping to along with the buyer assigned identifier code and the ship-to location. Some partnerships will require the N3 segment that will give the free form address that you shipping to. The N4 segment will also be required if the N3 segment is included. The N4 segment will give the City, State, and Zip Code information. Now the order level data will start. The next segment is the HL segment with an "O" in it, to show that this is the order document. This HL segment will increment as the second HL segment and now will have a numeric value to show that this is the first order segment. The next segment is the PRF segment. This segment is the purchase order number and the purchase order date. The next segment that is still part of the order level is the N1 segment with the buying information in it. This N1 segment will require the buyer assigned identifier and the store number in it. The next HL segment will begin the carton level data and will be represented with a "P" and a numeric number to show that it is the third HL segment and a numeric number to represent the fact that this the second HL segment in the Order Level. It may be important to know that the HL segments are known as the hierarchy segments. The next segment is the PO4 segment. This segment will specify the physical quantities, packaging, weights and dimensions relating to the items or styles. The next segment is the PKG segment. This is the outer container type. This segment has the item description type, which is represented with an "S" for structured. The next code is the VICS Code that is the association qualifier code, represented by the code VI. In this segment there is a packaging description code which has two parts to it. The first part is the container type, an "L" is the logical container and a "P" is the physical container. The second part of the packaging description code is represented with a numeric number such as an 01 for a carton, 02 is for a carton hanging garments, 03 is for a carton with hangers, 04 is for a carton with identifiable inner packs, 05 is for a carton with unidentifiable inner packs, 06 is for a rack with hanging garments. The next segment is the MAN segment that is represented with a "GM" code and a twenty digit carton number. This number will have the first four digits that will represent either cartons or pallets. The next six digits will be your UPC vendor code.

Your UPC vendor code is a number that will represent you company or company name within your company. This number is assigned by the Universal Council Code and kept on file to represent your company anytime it is used. Then you have nine digits that will be a sequential number or number that has no meaning except in this situation. A number that cannot be duplicated. The last digit is the twentieth digit that is the check digit. This number is created by a calculation that uses the first nineteen number of the twenty digit carton number. This number must be on your in you UCC 128 label. This number is how you shipment gets checked in at the distribution center or at the stores. In many cases, the ASN is also used to adjust inventory as it is loaded on to your partnerships system. The next segment is the next hierarchy level, also known as the next HL segment. This one is the start of the item or sku data level. This HL segment is represented with an "I" for item level. This segment also will increment by one for every item that is in this shipment and this item HL level will also represent the items that are in the carton that you have created the carton number for. This segment will also have the id number of the carton which is the pack number incremented by one. The second count of the HL hierarchy of the pack segment is the count of your cartons. The next segment will be the LIN segment. This segment will specify the Item or the styles that is being shipped with the UPC code that is represented for that item or style. The next segment is the SN1 segment. This segment will specify the quantities and unit of measure of the UPC code that is represented in the LIN segment.

The Item HL segment will be the start of the loop for many different items or styles that are in the carton. This loop will also have a LIN and SN1 segment in it. The final four segments are the CTT segment (control counter), the total number of detail lines, the SE segment is the partner of the ST segment, which is the transaction set trailer, the GE segment is the partner of the GS control number, which is the group control number, and the IEA segment is the partner of the ISA control number, interchange control trailer.

To review the hierarchy segments, better known as HL segments, there are four segments. First is the shipment level, which relate all of the trucking information, the name of the truck that will pickup and deliver the merchandise to the stores or distribution centers.

Included with this information are the Bill Of Lading, Ship Date, and Ship-to Information, such as the name of the partnership. The next HL segment is the order level, which will relate the order information, such as the purchase order number and the buying office information. The next segment is the shipment level, which will relate the shipment information, such as the sku quantities in the standard pack, outer container information, and the carton identification information.

The following is an example of An Advance Ship Notice:

ISA*00* *00* QL*S/R ID QL*S/R ID *Date *Time*US Code*Version*Control #*0*P (production code, T for test code)*>

GS*SH*S/R ID*S/R ID*Date*Time*Data Control#*ACK Request*Version

ST*Document number (856)*Transaction Control Number

BSN*the code to show the the status of the order (00=original ship notice, 07=duplicate ship notice)*ship notice number (usually the invoice number or a unique number assigned by the vendor or shipper)*ship notice date (usually the date that the ship notice was created)*ship notice time (hours, minutes, and seconds)*hierarchy structure code (this is the code indicating the packing structure. "0001" is for the pick and pack structure. "0002" is for the standard carton pack structure.

HL*hierarchical id number (this is the number that should be assigned sequentially within the transaction set.)*hierarchy level code (this code identifies the beginning of the shipment level data. The shipment level will only occur once.)*

TDS*routing sequence code (O=original carrier, m=motor carrier)* trucking identification code qualifier (usually a "2")*trucking identification code (SCAC code – abbreviation code that will represent the name of the trucking company.)

REF*reference number qualifier (BM = bill of lading)*bill of lading number (the bill of lading of the shipment)

DTM*time and date qualifier (011=the ship date)*schedule ship date (CCYYMMDD=century, year, month and day)

DTM*time and date qualifier (067=the cancel date)*scheduled cancel date (CCYYMMDD=century, year, month and day)

N1*entity identifier code (ST=ship-to location)*name of the ship-to location (name of the partnership location that the merchandise id being shipped to.)

Id code identified (buyer assigned – 92=store location)*Identification code (ship-to location Id=store or distribution number)

HL*hierarchical Id number (continuing sequentially number. There will be one order level data area for each different order in the shipment.)* hierarchical level code (O = order level)

PRF*retailer purchase order number (purchase order number)***purchase order date (date of the purchase order)

N1*entity identifier code (buying location – BY Identification code)*buying location name (name of the buying location or the name of the vendor)* Id code qualifier (92 = buyer assigned, buyer location)*Identification Code (store number)

HL*hierarchical Id number (incrementing sequential number, added on from the shipment and order hierarchy levels)*hierarchy parent Id number (Id number of the order level, which is related to the HL shipment segment.) hierarchical level code (P = pack level is for the carton data)

PO4**size of the item or style (usually not used for Pick/Pack structure quantity of the SKU within the pack for the standard pack structure.)* unit of measure (EA = eaches, CA = Cases for the standard pack structure)

PKG*Item description type (S = structured)*packaging character code (36 is a representation of the packaging specifications)*assoc iation qualifier code (VI = VICS is part of the version)*packaging description code (the code is broken down into two parts, the first part is the alpha, L = is a logical container to be used, P = is a physical container to be used. The second part in a numeric code to indicate the following)

01 = carton

02 = carton, hanging garments

03 = carton, with hangers, but not hanging items or styles

04 = carton, with the inner packs being identifiable

05 = carton, with the inner packs being unidentifiable

06 = rack, hanging garments

MAN*marks and numbers qualifier (GM = GMAIC format)*marks and numbers (GMAIC format carton ID number. This is the 20 digit carton number)

HL*hierarchical Id number (sequential Id number, adding to the order hierarchy Id number)*hierarchical parent Id number (Id number of the carton level. Sequential number adding one to the parent Id number in the pack level, but this number will not increment as long as you stay in the Item hierarchy level segments.*hierarchical level code (I = item level)

LIN**product Id qualifier code (UPC = UPC code that will identify how you will be identifying the item or style the is being shipped at this time)*the product Id (UPC code that will identify the item or styles that is shipped at this time.

SN1**total quantity (the total amount of pieces shipped for the above product code, if the next field is a code of EA. If the next field is a code of CA, then this will be the total quantity of the above case number)*unit of measure (EA, CA, DZ).

CTT*the total number of hierarchy (HL) segments in the transaction set* number of units shipped

SE*Number of Included Segments*Control Number

GE*Number of Included Transaction Sets*Data Interchange Control Number

IEA*Number of Included Groups*Interchange Control Number

Notes to Remember

The invoice document, also known as the 856 document, has a field separator which is created by binary coding, also known as a delimiter. Each segment has a record control delimiter, which is also created by binary coding.

Always keep in mind that your partnership may post in their documentation what delimiters that they want you to use. This means that you will have to include these delimiters in your document so their system will be able to read the document.

The hierarchy segments may need more explanation then the other segments in the ASN because they control the different parts of the order being shipped. As in any order or invoice that is created, you will have a section for your header information, Such as your shipping and billing information. Then there are the references to the order, such as your Bill of Lading, your dates, your trucking, and your total number of cartons that you will be shipping. In the detail of the order or invoice, you will have your style and quantity information broken down. In your ASN, your hierarchy (HL) segments do this for you. Your first hierarchy segment will start with an "S", for shipment. Always keep in mind that the HL segments are numerically sequenced. The shipment segment will start with a one, but no secondary level has occurred at this time. (HL*1**S). The next hierarchy segment is the HL segment for the order section. In this section you will reference the purchase order number, packaging information, shipping dates, department that your merchandise is going to, and in some cases, your internal vendor number. The first field is the numeric count of the hierarchy order segment. This will be a two, because this is the second hierarchy segment in the document. (HL*2*1*O). The second field will be a numeric one, because it is the first hierarchy order segment. The next hierarchy segment is the pack segment. The pack segment is were your pack or better known as your carton number is.

This number will be scanned in and will come up on the screen, and show all of your items or styles that are in your carton. The hierarchy pack segment is your third hierarchy segment to appear in this document. The first field will be a numeric value of three, because it is your third hierarchy segment in this document. The second field is the numeric value of two, because the pack relates back to the order segment which is a numeric one. (HL*3*2*P).

In most cases the value of the pack will remain the same throughout the document, but the value of the first field will continue to increment because you are adding more packs to this document. Your next pack or second carton will look like this. (HL*4*2*P). After you have given your pack or carton number, your next and last hierarchy segment is the hierarchy item segment. The numeric value of the first field will continue on from the first value in the first field of the pack segment. The second field will also continue on from the numeric value of the second field in the pack segment. (HL*4*3*I) and will keep incrementing for each item or style that is in each case.

7

Creating and Using the UCC 128 Label

The UCC 128 Label is a carton label that will represent the electronically transmitted Advanced Ship Notice. This label has many versions and still is very similar. The common factor in this label is the UCC 128 carton number. This is the twenty digit number that will be represented by a bar code. This bar code will be read on the warehouse belt as the merchandise rides along. When this bar code is read, the matching number in the ASN will bring the information to the screen and make you aware of the contents in the carton along with the other existing information. The partnership that you are creating this label for will have their own specifications to follow when using this label.

The warehouse has the job of making sure that the label is in the proper place on the carton. This is important to the distribution center or processing center. As the cartons are taken off of the truck, they are put on a conveyor belt. This belt is moving and the cartons will ride under the scanner. The scanner usually does not cover the whole box, so this why it is important to have the UCC 128 label put on the box according to the partnerships guide lines.

The UCC 128 label is important for the trucker and the warehouse to direct the cartons to the next freight carrier to get merchandise to the proper store where the merchandise will be off loaded and brought in to be stocked on the shelves. That is the use for the label in general terms. This part does not have to do with the scanning of the label.

The UCC 128 label for scanning purposes starts in the warehouse. The UCC 128 label is four inches across and six inches long.

This label will have your Name and Address on the top of the label. Then the partnerships Name and Address will also be on the label. Some partnerships have their Name and Address on the top of the label and some have it under your Name and Address. This label is four inches across and six inches in length.

The Following is an example of a UCC 128 label:

```
FROM:            !    To:
XXXNAMEXXX       !    XXXXNAMEXX
XXADDRESSXX      !    XXADDRESSXX
CITY, STATE, ZIP !    CITY, STATE, ZIP
_____
SHIP TO POST     !  BILL OF LADING
(420) 31025      !  PRO NUMBER
    !!!!!!!!!!!!!!!!!!   !
    ! BAR CODE!!!   !
    !!!!!!!!!!!!!!!!!!   !  CTN 1 OF 1
_____
PO #:  1234567890      STORE #
DEPARTMENT: 123    0014
_____
        00 00 123456  7890123456

    !!!!!!!!!!!!!!!!!!!!!!!!!!!!!!!!!!!!!!!!!!!!!
    !!!!!!!!!!!BAR CODE!!!!!!!!!!!!!!!!!!!!!
    !!!!!!!!!!!CARTON NUMBER!!!!!!!!!
    !!!!!!!!!!!!!!!!!!!!!!!!!!!!!!!!!!!!!!!!!!!!!
```

There are different designs of this label. The design will vary according to the partnership that you are working with. Some labels will have letters in two points, four points, 10 points, or 20 points. Points will determine the size of the letters on the UCC 128 label. Some fields such as the store number or the purchase order number, would be easier to read if the letters or number is shown in bigger letters or numbers.

Some labels may have additional fields on them. Some partnerships may require the overall description of your merchandise.

Some partnerships will even have a chart to follow or table up with the set of descriptions that they have. Some partnerships will require a UPC Code or an item number or style number. Some partnerships will require a bar Code with the store number. The label always will allow a little extra room for some extra fields that you or your warehouse may need to put the labels on the cartons. These fields may be the style number or item number, carton numbers or the description of the style or item. Your postal Code will require a code of "420" in the bar code along with the zip code. All bar codes will require a quiet zone before the bar code is printer for the scanner in the warehouse. When the scanner starts to move over on to the bar code, it will move into the start of the bar code. If the quiet zone is not there, when the scanner starts, it would cut off the beginning of the number. This quiet zone is usually a quarter on an inch. This will apply for all of the bar codes on the label. The usual size of the UCC 128 bar code is 1 ¼ inches high and 3.02 inches wide.

8

Utilizing Other Documents

All partnerships are using or starting to use the other

documents that have been created to keep the flow of information connected with the original setup of EDI documents. These documents are all stand alone documents put all have ties to the other documents that were sent.

We will explore the other documents and how they can work for you. The first document is the (820) document. This is the Remittance Advise Document. Many partnerships have gotten away from sending statements and advice sheets with your checks. This document will detail your check or checks by giving you a check number, a check amount, a check date, a code to represent to transaction as a debit or credit. The check amount will be a negative if the amount is a credit. The can be other transactions that can be represented in this document. There are charge backs, discounts, allowances and freight. This will be determined by the transaction codes that are sent in the document.

This document will start off as all other documents start with the opening of the envelope.

The Remittance Advice begins with same segments as the other documents with the ISA envelope opener. The ISA will carry the Sender/Receiver number of your company and the Sender/Receiver number of your partnership. This information would be posted on your partnerships Web-Site. In many cases, you will find that the Sender/Receiver ids will not be the same as the other documents.

This allows the partnership to have the extra separation when processing the document on their system. The next segment is the GS segment. The GS segment is the same as the other documents except the Sender/Receiver numbers again will vary from partnership to partnership. This GS segment will have a code of RA in it to show that this is the Remittance Advice document.

The next segment is the ST segment. In this segment, the document number will be in the first field as an 820. The next segment, which is the BPR segment, you will start the summary of the remittance being sent to you.

In the first field, you will state the transaction code, usually you get three choices, Remittance, Payment with remittance advice, or only a Payment, next will be the total amount of the check. Continuing on you will be confronted with codes to justify the actions of your partnership. Is it a Debit or a Credit, what payment code is used for payment. Was the payment wired? to your account. Was the payment deposited into your account directly. The check date or the deposit date. If the amount was directly deposited to your account, or wired into your account, then they will need your bank account number. The next segment would be dealing with the tracing number that will be sent to you. This segment is the TRN segment. Starting with the tracing code and followed by the check number to be traced. The next segment will be the REF segment. This reference is the batch number assigned to the whole transaction. The next segment is the DTM segment.

In this segment you will be given the transmission date. The next segment, you have the N1 segments that will tell you your partnerships phone number or your partnerships duns number. Following that N1 segment, with another N1 segment that will state your phone number or duns number. The next segment is the ENT segment. In this segment, your partnership will identify themselves by their name and also with a buyer assigned number. Followed by the NM1 segment. This will define any sub-divisions that you partnership will have with another buyers assigned number. Anytime you see a buyers assigned number, this could be the phone number, duns number or your store number.

Now we have moved into the summary segments. We will start with the adjustments. In this ADX section, they will give you the adjustment amount. This amount will be a positive amount if it is a debit and a negative amount if it is a credit. This amount is followed by the debit code or the credit code. The next segment, the REF segment will tell you what department your partnership is referring to followed by the date this transaction was created in the DTM segment. The following segments will reference all of the purchase order and invoice information. The final segment will give you the final dollar amount of the check after all or any deductions were taken out. The SE segment is the partner of the ST segment, which is the transaction set trailer, the GE segment is the partner of the GS control number, which is the group control number, and the IEA segment is the partner of the ISA control number, interchange control trailer.

Notes to Remember

The invoice document, also known as the 820 document, has a field separator which is created by binary coding, also known as a delimiter. Each segment has a record control delimiter, which is also created by binary coding.

Always keep in mind that your partnership may post in their documentation what delimiters that they want you to use. This means that you will have to include these delimiters in your document so their system will be able to read the document.

The next document is the 852 document. This document is the Product Activity Data Document. In this document the activity of your sales are sent back. This is to let you know what products have been selling and the quantities that have sold and the stores that they sold in. This document will in some cases mirror the purchase order document by using the SDQ segment. This segment allows multiple stores and quantities to be sent at one time, in one segment. The XQ segment is setup for the purpose of showing the reporting date and the location. This next segment is the LIN segment. In this segment, your partnership will transmit your UPC code or if you use a European product code, they will transmit back the EAN code to represent your item code or style number. The UPC code is twelve digits and your EAN Code is thirteen digits. The next segment is the ZA (Product Activity Reporting) segment. In this segment your inventory status will be sent to you. You can review your current inventory quantities that would be available for selling or for a shipment. In addition you will also be told of any additional demand quantities. The next segment will be the actual quantity. The next segment will be the unit of measure and then the date that this information will relate to. The LIN and ZA segments can loop for as many items or styles that your partnership may be sending back to you. The SE segment is the partner of the ST segment, which is the transaction set trailer, the GE segment is the partner of the GS control number, which is the group control number, and the IEA segment is the partner of the ISA control number, interchange control trailer.

The following is an example of the Product Activity Report:

ISA*00* *00* QL*S/R ID QL*S/R ID *Date *Time*US Code*Version*Control #*0*P (production code, T for test code)*>

GS*PD*S/R ID*S/R ID*Date*Time*Data Control#*ACK Request*Version

ST*Document number (852)*Transaction Control Number

XQ*H (handling code, this is for notification only)*CCYYMMDD (date of the report results that are being sent to you)*RL (reporting location code)*

reporting location code (free form description of the reporting location code)

LIN*assigned identification number (this will be sequentially assigned to keep track of the number of lines*UP or EN (represents the UPC Code or the EAN Number that will follow in the next field)*the UPC Code or the EAN Number

ZA*QA or QD (QA is the code that will represent the current quantity that is available in your inventory that is for a shipment or for a sale. QD is the code that will represent any additional quantity for demand that is over and above the normal amount for the replenishment calculation)*Quantity (the actual amount of pieces that your partnership is referring to)*EA (the unit of measure. EA = Eaches, CA = Cases)*007 (code to represent the following segment that will be the date of when the inventory has been taken)*CCYYMMDD (The actual date that these quantities are as of)

CTT*the total number of segments in the transaction set*number of units shipped

SE*Number of Included Segments*Control Number

GE*Number of Included Transaction Sets*Data Interchange Control Number

IEA*Number of Included Groups*Interchange Control Number

Notes to Remember

The Product Activity Data document, also known as the 852 document, has a field separator which is created by binary coding, also known as a delimiter. Each segment has a record control delimiter, which is also created by binary coding.

Always keep in mind that your partnership may post in their documentation what delimiters that they want you to use. This means that you will have to include these delimiters in your document so their system will be able to read the document.

The next report is the Purchase Order Status Report. This report is also known as the 870 Order Status Report. This report can be used to report on the current status of a purchase order or on selected lines of the purchase order for purposes of updating any of your scheduled shipments and your delivery dates of those shipments. The 870, Order Status Report will start the same as all of the other documents. The ISA will carry the Sender/Receiver number of your company and the Sender/Receiver number of your partnership. This information would be posted on your partnerships Web-Site. In many cases, you will find that the Sender/Receiver ids will not be the same as the other documents. This allows the partnership to have the extra separation when processing the document on their system. The next segment is the GS segment. The GS segment is the same as the other documents except the Sender/Receiver numbers again will vary from partnership to partnership. This GS segment will have a code of RS in it to show that this is the Order Status Report. Then the ST segment will also have the code 870 in it along with the sequential number that will be paired up with the SE segment. The BSR segment is the segment that will begin the order status report. The BSR segment is coded with specific codes to represent the type of report this document is going to produce. In this segment you will also have your unique number and purchase order date.

The next segments will be the REF segments that will usually refer to the internal vendor code and the department number.

As you may recall, in the Advance Ship Notice there were hierarchy segments that represented the four levels of the shipment. In the Order Status Report there are two levels of hierarchy levels. The first one will be the "O" order level. In the order level, you will start to bring in the detail of the purchase order that you are sending over. This PRF segment is the next on to follow the HL order segment. The PRF segment will bring in the purchase order number and the purchase order date. The next segment to follow is the N1 segment with an ST code to represent the store number that the order is being shipped to. The next segment will bring in the second of two hierarchies levels. This HL segment will be the "I" Item level. The first segment in the HL Item level is the PO1 segment. The PO1 segment is very similar to the PO1 segment that was brought in through your "850" Purchase Order Document. You will have you line count, because you can have multiple lines, such as your purchase order has, total quantity and unit of measure.

The unit of measure can be represented in EA for Eaches, CA for Cases, or DZ for dozens. In many cases you may be sent a CB for the catalog number and or the UPC code that is cross referenced on the purchase order.

This will depend on the setup you have with your partnership. This segment is followed by the ISR segment. The ISR segment is the item status of the purchase order. This segment is usually coded with a CP, for partial shipment with no back order, or an IC, for an item cancelled. If the code is CP, then the next field will be the quantity of the partial shipment.

CTT*the total number of hierarchy (HL) segments in the transaction set* number of units shipped

SE*Number of Included Segments*Control Number

GE*Number of Included Transaction Sets*Data Interchange Control Number

IEA*Number of Included Groups*Interchange Control Number

Notes to Remember

The Product Activity Data document, also known as the 870 document, has a field separator which is created by binary coding, also known as a delimiter. Each segment has a record control delimiter, which is also created by binary coding.

Always keep in mind that your partnership may post in their documentation what delimiters that they want you to use. This means that you will have to include these delimiters in your document so their system will be able to read the document.

The following is an example of an Order Status Report:

ISA*00* *00* QL*S/R ID QL*S/R ID *Date *Time*US Code*Version*Control #*0*P (production code, T for test code)*>

GS*RS*S/R ID*S/R ID*Date*Time*Data Control#*ACK Request*Version

ST*Document number (870)*Transaction Control Number

BSR*3 (unsolicited report)*PP (selected orders-selected items)*(completely unique number)*CCYYMMDD (order status report creation date)

REF*IA (your internal vendor code)*your internal vendor code

REF*DP (department code)*your department number

HL*1**O (first hierarchy segment, to show that you are now in the order section of the document

PRF*purchase order number***purchase order date

N1*ST (ship-to information)**92 (buyers assigned code)*ship-to location (store number or distribution center)

HL*2*1*I (second hierarchy segment, to show you that the first item is going to follow)

PO1*1 (first line of item or style detail)*quantity*unit of measure (ea = eaches, ca = cases, dz = dozens)***CB (code for buyers catalog or UP for UPC code)*catalog number or UPC code

ISR*CP (Partial shipment or IC for item cancelled)*CCYYMMDD (the date of the partial shipment or the date that the item was cancelled QTY*01 (code for a discrete quantity)*quantity (quantity that was

shipped or cancelled)*unit of measure (ea = eaches, ca = cases, dz = dozens) (The HL segments thru the QTY segment can be repeated as many for as many items that are in the purchase order. The HL numbers will increase by one to show that the HL will be the next HL and to show that the HL is representing the next item)

CTT*the total number of hierarchy (HL) segments in the transaction set*number of units shipped

SE*Number of Included Segments*Control Number

GE*Number of Included Transaction Sets*Data Interchange Control Number

IEA*Number of Included Groups*Interchange Control Number

The next document that id commonly use is the Text Document. This document is known as the 864 document. The Text document is used for the purpose of letting the vendor know that there was a problem with a certain document that was sent. As a reminder to you when you do receive this document, you must remember to send a functional acknowledgement back to your partnership. The reason is that the information in this document will not be sent to you in any other way. The main use of this document is thru invoicing. The Text Document begins with same segments as the other documents with the ISA envelope opener. The ISA will carry the Sender/Receiver number of your company and the Sender/Receiver number of your partnership. This information would be posted on your partnerships Web-Site. In many cases, you will find that the Sender/Receiver ids will not be the same as the other documents. This allows the partnership to have the extra separation when processing the document on their system. The next segment is the GS segment. The GS segment is the same as the other documents except the Sender/Receiver numbers again will vary from partnership to partnership. This GS segment will have a code of TX in it to show that this is a Text document. The next segment is the ST segment. In this segment, the document number will be in the first field as an 864. The BMG segment is the next segment. This is the beginning of Text messaging. In this segment, you will have the original code. This code will denote that this is the first time that this message is being sent. That will be followed by a description that will clarify the data that is being sent to you. This is then followed by a report message code that will state that the report is to be printed. The next segment will be the DTM segment, which will show the creation date of the text message. The next segment is the N1 segment.

This will state who the text message is from, the store name, duns number, and store number. The next segment is to tell you your vendor's name. Along with your vendors name, the next segments will be REF segment that will state your internal vendor code. This is now followed by the MIT segment, which is the message identification segment. This will tell you what the message would like you to do. Then the next segment is the MSG segment that will tell you what the problem is. The actual message will almost always be the reason why this Text Document was sent. The SE segment is the partner of the ST segment, which is the transaction set trailer, the GE segment is the partner of the GS control number, which is the group control number, and the IEA segment is the partner of the ISA control number, interchange control trailer.

The following is an example of a Text Document:

ISA*00* *00* QL*S/R ID QL*S/R ID *Date *Time*US Code*Version*Control #*0*P (production code, T for test code)*>

GS*TX*S/R ID*S/R ID*Date*Time*Data Control#*ACK Request*Version

ST*Document number (864)*Transaction Control Number

BMG*Original Code (00)* Free Form description to clarify data*report message

DTM*097 (date creation code)*CCYYMMDD (creation date of the text Message)

N1*FR (From the store related to your partnership)*9 (buyers assigned number)*Duns number plus the store number

N1*TO (to your company name)*Your company name

REF*IV (internal vendor number code)*Internal vendor number

MIT**What actions you are to take upon reading this message

MSG*this message this the related situation

SE*Number of Included Segments*Control Number

GE*Number of Included Transaction Sets*Data Interchange Control Number

IEA*Number of Included Groups*Interchange Control Number

Notes to Remember

The Product Activity Data document, also known as the 864 document, has a field separator which is created by binary coding, also known as a delimiter. Each segment has a record control delimiter, which is also created by binary coding.

Always keep in mind that your partnership may post in their documentation what delimiters that they want you to use. This means that you will have to include these delimiters in your document so their system will be able to read the document.

The routing guides have become an important part of our shipping to certain stores. The routing documents are the 753 and 754 documents. These documents are used to inform the partnership and the trucking company that there is a shipment scheduled and the information in the 753 document are all of the details of the shipment. The information will include the invoice number, ship date, weight, cubic inches, and bill of lading. As you transmit this information over to the sender/receiver Id's that your partnership has given to you, when your 753 document has been processed, they will return a 754 document to you. In the 754 routing document, you will find a reply to your transmitted 753 document. This document will have the trucking company that will be picking up your merchandise and you will be given bill of lading number that they will want you to use in your Advance Ship Notice. After you receive the 754 shipping document, you must remember to send back a functional acknowledgement.

The follow is an example of a 753 routing document:

ISA*00* *00* QL*S/R ID QL*S/R ID *Date *Time*US Code*Version*Control #*0*P (production code, T for test code)*>

GS*RF*S/R ID*S/R ID*Date*Time*Data Control#*ACK Request*Version

ST*Document number (753)*Transaction Control Number

BGN*Original or Duplicate Codes (00 or 07)*Invoice number (a unique number to represent the shipment)*CCYYMMDD (Creation Date of this routing document)*HHMM (the time of the creation of this routing document)

PER*Reporting information code****E-mail or telephone code*E-mail address or telephone number

N1*SF (ship from code)*Your vendor name*92 (buyers assigned code, could also be 9 or 91)*Duns number or telephone number

N3*Your vendors free form address

N4*Your vendors city*your vendors state*Your vendors zip code

LX*Invoice Number

N1*ST (ship to code)**92 (buyers assigned code)*Partnerships store number

L11*Partnerships reference number (can use your UCC 128 carton number. the last digit (20th Digit), must be a sequential number that is added by one for every invoice of this shipment, that is processed)*RRC (routing code)

G62*RS (routing for shipment code)*CCYYMMDD (Date of the shipment that you are routing)*RS (again the routing for shipment code)*HHMM (Time of shipment that you are routing)

OID**Purchase order number of shipment*107 (standard code)*CTN (carton code for the number of cartons, that will follow)*Number of cartons that will shipped with this purchase order number and for this invoice number)*L (code for weight of the shipment)*Weight of the merchandise of the shipment for this invoice)*E (cubic feet code)*Cubic feet of the shipment (this is measured with one decimal)

OID**Purchase order number of shipment*107 (standard code)*PCS (pieces or units code for the number of cartons, which will follow)*Number of pieces or units that will shipped with this purchase order number and for this invoice number)*L (code for weight of the shipment)*Weight of the merchandise of the shipment for this invoice)*E (cubic feet code)*Cubic feet of the shipment (this is measured with one decimal)

***** note that the first OID (order invoice detail is for the cartons and the second OID (order invoice detail is for the pieces or units)

SE*Number of Included Segments*Control Number

GE*Number of Included Transaction Sets*Data Interchange Control Number

IEA*Number of Included Groups*Interchange Control Number

Notes to Remember

The Routing document, also known as the 753 document, has a field separator which is created by binary coding, also known as a delimiter. Each segment has a record control delimiter, which is also created by binary coding.

Always keep in mind that your partnership may post in their documentation what delimiters that they want you to use. This means that you will have to include these delimiters in your document so their system will be able to read the document.

The 754 routing document is the answer to the 753 routing document. When your partnership receives the 753 routing document, they will get the information about the shipment and in the form of a 754 document, they will transmit it back to you. The documents usually have a forty eight hour window. Some of the information may seem to be the same, but they are just verifying the information. In most cases the bill of lading number is the number that is needed in your advance ship notice and to verify the trucking company.

The following is an example of a 754 document:

ISA*00* *00* QL*S/R ID QL*S/R ID *Date *Time*US Code*Version*Control #*0*P (production code, T for test code)*>

GS*RF*S/R ID*S/R ID*Date*Time*Data Control#*ACK Request*Version

ST*Document number (754)*Transaction Control Number

BGN*Original or Duplicate Codes (00 or 07)*Reference Id (a unique number to represent the shipment)*CCYYMMDD (Creation Date of this routing document)*HHMM (the time of the creation of this routing document)***Response Code

PER*Contact Function code*Name***E-mail or telephone code*E-mail address or telephone number

N1*SF (ship from code)*Your vendor name*92 (buyers assigned code, could also be 9 or 91)*Duns number or telephone number

N3*Your vendors free form address

N4*Your vendors city*your vendors state*Your vendors zip code

LX*Invoice Number (assigned number)

N1*ST (ship to code)**92 (buyers assigned code)*Partnerships store number

L11*Partnerships reference number (can use your UCC 128 carton number. the last digit (20th Digit), must be a sequential number that is added by one for every invoice of this shipment, that is processed)*RRC (routing code) BLR*Trucking Alpha Code (routing for trucking code)

OID**Purchase order number of shipment*********Code to verify invalid quantity, store number, purchase order number, or outside the shipping window.

G62**Pickup Date (CCYYMMDD)**Pickup Time (HHMM)

MSI*Multi Stop Shipment

QTY**Quantity to be shipped

N1****Ship to Name (partnership store)

SE*Number of Included Segments*Control Number

GE*Number of Included Transaction Sets*Data Interchange Control Number

IEA*Number of Included Groups*Interchange Control Number

Notes to Remember

The Routing document, also known as the 754 document, has a field separator which is created by binary coding, also known as a delimiter. Each segment has a record control delimiter, which is also created by binary coding.

Always keep in mind that your partnership may post in their documentation what delimiters that they want you to use. This means that you will have to include these delimiters in your document so their system will be able to read the document.

9

How This Works For You

As we have reviewed the documents and all of the information that you as a vendor will need to provide to your partnerships, and all of the information that your partnerships will provide you with, you would have to wonder how all of this will work for you.

We can start with the purchase order that is being transmitted to your vendor. If you have a system like my system, then you will not have to do any manual entry, because the transmitted information will be brought into your system as an open order and will remain there until the purchase order is picked and billed. When the information is billed, the information will then be sent out to the invoicing and advance ship notice holding path. At this time you should be able to transmit the invoices and the advance ship notice. The UCC 128 labels should be created at this time and sent to the warehouse so they can put them on the cartons.

When these cartons arrive at the distribution centers or directly to the stores, they will be put on a conveyer belt, in some cases the conveyer belt is extended into the trucks so the cartons do not have to be lifted out of the truck. Then when the carton travel down the conveyer belt, they will pass under the scanner and be accounted for thru your transmission that was sent to your partnership. Then the cartons will continue on and be loaded back on the truck, if you are at a distribution center, and this truck will be the truck that brings the cartons to the stores. If they are scanned in at the Store level, then the cartons will go directly to be they can start putting them on the shelves. When these cartons ride the conveyer belt, it is important that the label is in the proper area on the carton, so the carton can be read in.

This overall procedure at the distribution center and the store level will save a lot of time for you because the manual counting usually takes days to finish. What manual counting means is that your partnership will have to pick up the cartons. Then they will have to open it and start to count your merchandise. If there are multiple cartons, then they have to keep track of all your items or styles. Then they will need to get your original purchase order and check in there counts against the purchase order.

To review the time saving possibilities:

1) Transmitted purchase order – a chance to omit manual entry and keying in errors.

2) Sending functional acknowledgements

3) Transmitting invoices – no time delay for mail or manual entry after they receive them. A chance to omit keying in errors.

4) Receiving a functional acknowledgement as receipt of your invoices

5) Transmitting Advance Ship Notice – Your shipment gets checked in immediately and your cartons do not have to be opened and will get loaded onto the proper truck that will deliver the all of your cartons to the proper stores.

6) Receiving a functional acknowledgement as a receipt of you advance ship notice. With some partnerships the advance ship notice is used as the proof of delivery.

7) Remittance Transmissions – transmitted into your system. You can then print out a report or update your system.

As you know, all of these documents need to be correct when they are sent to your partnership. The people who are working to make these documents happen must know all of the details at all times. If the information that is transmitted, is correct, then the flow of this process will go smooth. If you partnership detects any problems at any time, and it is detected in a timely manner, then the error usually can be corrected and resent in time for your transmission to work as it is intended to work.

About The Author

My name is Alan Castro. I have been a computer consultant for the past twenty five years. In the 1980's, I was apart of the process of developing the Electronic Data Interchange, also known as EDI. I had put the first company in use with EDI. As the years went on I have developed EDI to flow thru the software packages so there would be no manual entry.

I have also developed an EDI package that can run on its own. This is when the data is transmitted into your system, the data is also ready to be transmitted as an invoice or an advanced ship notice.

I have been setting this up thru my company Unique Software Support. Please write me at usscorp13@aol.com.